Green legacy

Leading the way to a sustainable planet

By

Gary D. Gates

Copyright

Copyright © Gary D. Gates 2024, All rights reserved. This document and its contents are protected under international copyright laws. No part of this document may be reproduced, distributed, or transmitted in any form or by any means, including photocopying, recording, or other electronic or mechanical methods, without the prior written permission of the copyright holder, except in the case of brief quotations embodied in critical reviews and certain other noncommercial uses permitted by copyright law.

Table of content

Copyright

Table of content

Chapter 1 The State of Our Planet

Chapter 2 Redefining Sustainability

Chapter 3 Renewable Energy and the Way Forward.

Chapter 4 Transforming Our Food Systems

Chapter 5 Revolutionizing Transportation

Chapter 6 Responding to the Global Water Crisis.

Chapter7 Green Jobs and the New Economy.

Conclusion Maintaining a Green Legacy

Introduction

The need for sustainable solutions has never been more pressing than it is in this day and age, when climate change, environmental degradation, and the depletion of resources are the most prominent topics in the news. There is a stark choice that must be made in light of the current trajectory of the health of our planet: either continue on the path of activities that are not sustainable or pivot towards a future in which humans live in harmony with other living things around them. Through the use of motivation, education, and empowerment, the book "Green Legacy: Leading the Way to a

Sustainable Planet" intends to encourage individuals and communities to pursue the latter road. This book is not merely a rallying cry; rather, it is an all-encompassing handbook that explains how we may collaboratively work toward making the world a more sustainable place.

The concept of sustainability is no longer a specialized concern that is just present among scientists or environmentalists; rather, it has evolved into a worldwide necessity that permeates every facet of our existence. The decisions that we make affect the environment in every way, from the food that we eat and the energy that we use to the items that we purchase and the mode of transportation that we employ. This book is significant because it confronts these problems head-on, providing solutions that are both practical and inspiring examples of how we may make a difference in the world with our actions.

The idea of sustainability incorporates not only the state of the environment but also the state of the economy and the distribution of resources about social equality. A sustainable world is one in which resources are utilized effectively, waste is reduced to a minimum, and everyone has access to clean air, water, and energy throughout the entire planet. Through the adoption of sustainable practices, we are not only preserving the earth for the benefit of future generations, but we are also establishing a society that is more just and just. This book acts as a guide to help readers navigate these complicated concerns, providing them with the information and tools they need to contribute to a sustainable future.

A Crucial Moment in the Course of History

We are currently living in a period that will be remembered by future generations as either the moment when we were able to turn things around or the moment when we were unable to take advantage of our final opportunity. The scientific community has reached a unanimous agreement that human actions are the primary cause of climate change. This change is causing weather events to become more frequent and severe, as well as rising sea levels and a loss of biodiversity. We have a short window of opportunity to make significant adjustments to avoid catastrophic repercussions, according to the Intergovernmental Panel on Climate Change, which warns as much.

The current critical moment is characterized by both enormous problems and opportunities that have never been seen before. There are several formidable problems, including the reduction of greenhouse gas emissions, the shift to renewable energy, the protection of endangered species, and the guarantee of sustainable resource utilization. The prospects, on the other hand, are similarly extensive. There is a one-of-a-kind opportunity to adopt reforms that can lead to a society that is more resilient and sustainable, and this opportunity is made possible by technological developments, legislative shifts, and increased public awareness.

This is a time of rapid innovation and transformation. Renewable energy technologies, such as solar and wind power, have become more economical and efficient, making them viable alternatives to fossil fuels. Advances in energy storage, such as battery technologies, are solving the intermittency difficulties with renewable energy sources. Additionally, new agricultural approaches are emerging that can feed the expanding global population without destroying the planet's resources.

Furthermore, this time in history is characterized by a groundswell of grassroots movements and young involvement. From Greta Thunberg's Fridays for Future strikes to the global Extinction Rebellion marches, people throughout the world are demanding action on climate change. This collective awakening implies a shift in cultural ideals towards increased environmental understanding and accountability.

However, this important period also requires strong leadership and coordinated initiatives at all levels. Governments, businesses, and individuals must work together to enact policies and practices that support sustainability. International cooperation is necessary to address global concerns such as climate change and biodiversity loss. This book underscores the value of these relationships and presents examples of successful programs that may be scaled and duplicated.

The Power of One: How Individuals Can Drive Change.

While structural change is vital, the power of individual efforts should not be underestimated. Each person's choices, activities, and advocacy efforts contribute to the wider trend towards sustainability. This book highlights that everyone has a role to play and that little, everyday activities can together lead to large positive consequences.

Individual actions can take numerous forms. Reducing energy use, choosing sustainable products, limiting trash, and promoting renewable energy are just a few examples. By making intentional decisions, individuals may reduce their carbon footprint and inspire others to do the same. This ripple effect can lead to extensive changes in consumer patterns and societal norms.

One of the most potent ways individuals can drive change is via activism and education. By increasing awareness about environmental challenges and supporting sustainable practices, individuals can influence their communities and policymakers. Grassroots movements and community projects frequently start with a single person's vision and tenacity. For example, Wangar Green

Belt Movement began with her efforts to plant trees in Kenya and developed into a global environmental campaign.

Moreover, individuals may create change by holding corporations and governments accountable. Consumer decisions can force corporations to adopt more sustainable practices. Boycotting items that hurt the environment and supporting those that value sustainability sends a clear message to corporations. Additionally, voting for political leaders who prioritize environmental policy and participating in public demonstrations might influence government action.

Chapter 1
The State of Our Planet: An Overview

Our planet is a complex and dynamic system that supports a wide range of life forms through complicated interdependences. However, human activities are putting the system's balance at risk. As we approach a watershed moment in history, we must grasp the current state of our planet to successfully confront the issues that lie ahead.

Climate Change.

Climate change is today's most pressing environmental issue. The scientific agreement is that human actions, mainly the use of fossil fuels and deforestation, have resulted in an unprecedented increase in greenhouse gases (GHGs) in the atmosphere. The principal greenhouse gases that contribute to global warming are carbon dioxide (CO_2), methane (CH_4), and nitrous oxide (N_2O). According to the Intergovernmental Panel on Climate Change (IPCC), the Earth's average surface temperature has risen by around 1.2°C since pre-industrial times, with serious effects.

We are seeing increasingly frequent and severe weather disasters including storms, droughts, heatwaves, and flooding. These disasters not only produce immediate devastation but also affect ecosystems and human communities, resulting in long-term socioeconomic issues. Long-term droughts, for example, can damage agricultural production, resulting in food poverty and economic instability, whereas rising sea levels threaten to submerge coastal cities and displace millions of people.

Biodiversity Loss

The health of our ecosystems is inextricably linked to biodiversity—the range of life forms on Earth. Biodiversity supports ecosystem resilience, allowing natural systems to adjust to changes while still providing critical services like clean air, water, and fertile soil. Unfortunately, we are currently witnessing the sixth mass extinction, with species vanishing at an alarming rate as a result of habitat destruction, pollution, climate change, and overfishing.

The loss of biodiversity has far-reaching consequences for ecosystems and human well-being. For example, the reduction of pollinators such as bees and butterflies can have a significant impact on food production because many crops rely on them for pollination. Similarly, the degradation of wetlands and forests lowers natural water filtration and carbon sequestration capabilities, heightening concerns about water quality and climate change.

Resource depletion

Human activities are depleting natural resources at unsustainable rates. Non-renewable resources, such as fossil fuels, minerals, and metals, are being depleted faster than they can be replaced. Even renewable resources such as freshwater, forests, and fish stocks are under significant strain owing to misuse and mismanagement.

The depletion of resources not only endangers environmental stability but also offers substantial economic and societal issues. For example, water scarcity impacts billions of people worldwide, limiting access to safe drinking water and reducing agricultural production. Similarly, deforestation contributes to climate change, changes local weather patterns, and uproots indigenous groups that rely on trees for a living.

Carbon footprints and ecological imprints

To comprehend human impact on the environment, we must consider two fundamental concepts: carbon footprints and ecological imprints. These indicators help to assess the level of human effect on the environment and identify areas where we might lessen our impact.

Carbon Footprints

A carbon footprint is the total amount of greenhouse gases emitted directly or indirectly by human activity, expressed in equivalent tons of CO_2. This measure includes a variety of sources of emissions, such as energy consumption, transportation, food production, and waste disposal.

- **Energy Consumption:** The energy sector is the major source of global GHG emissions, mostly from the combustion of fossil fuels for power and heat. Transitioning to renewable energy sources, such as solar, wind, and hydropower, is critical for lowering carbon emissions.

- **Transportation:** Cars, trucks, planes, and ships are all significant sources of emissions. Promoting public transit, electric automobiles, and alternative fuels can assist in reducing this impact.

- **Food Production:** Agriculture and food systems contribute to emissions via deforestation, methane emissions from animals, and the use of synthetic fertilizers. Adopting sustainable farming practices and limiting meat intake can help to reduce our diet's carbon footprint.

- **Waste Management:** Landfills and incineration produce methane and CO_2, respectively. Enhancing recycling systems, composting organic waste, and minimizing single-use plastics are all efficient ways to reduce waste-related emissions.

Individuals and businesses may reduce GHG emissions by calculating and understanding their carbon footprints. Simple adjustments, such as adopting energy-efficient appliances, carpooling, and supporting renewable energy efforts, can all have a major impact.

Ecological imprints
While carbon footprints focus on greenhouse gas emissions, ecological imprints provide a more

comprehensive assessment of human impact on the ecosystem. An ecological footprint is a measure of the quantity of land and water required to produce the resources we use and absorb the garbage we generate. This measure includes a variety of aspects, such as food production, energy use, transportation, and waste management.

- **Bio capacity:** The ecological footprint is compared to the Earth's biocapacity, or ecosystems' ability to replenish resources and absorb waste. When the ecological footprint exceeds biocapacity, it signifies ecological overshoot, which means that humans are using resources faster than the world can renew them.

- **Consumption Patterns:** Different consumption patterns leave diverse ecological footprints. Diets strong in meat and dairy, for example, have greater ecological footprints than plant-based diets because livestock production requires a lot of space and water. Similarly, the production and disposal of electronic gadgets have substantial environmental consequences due to rare material extraction and the formation of e-waste.

- **Sustainable Practices:** Reducing ecological footprints entails implementing sustainable practices such as using renewable resources, reducing waste, and protecting natural areas. Initiatives such as sustainable agriculture, forest management, and water conservation are critical to preserving the balance between human activity and the planet's ability to support life.

Understanding ecological imprints allows us to grasp the big picture of our environmental influence and devise comprehensive solutions for living within the Earth's resources. By matching our consumption habits with sustainable practices, we may lessen our environmental footprints and achieve a more harmonious relationship with nature.

The Hidden Costs of Consumption. Modern consumer culture is distinguished by its emphasis on convenience, affordability, and rapid pleasure. However, this culture frequently obscures the underlying costs of consumption, which go far beyond the price tags of things. The hidden costs of consumerism include environmental damage, social inequities, and health problems that are not immediately obvious but have far-reaching long-term implications.

Environmental Degradation.

The creation, transportation, and disposal of consumer products all contribute considerably to environmental damage. These hidden expenses are frequently not represented in the pricing of products and manifest in a variety of ways:

- **Resource Extraction:** The extraction of raw materials like metals, minerals, and fossil fuels frequently destroys habitats, causes soil erosion, and pollutes water. For example, mining operations can pollute water sources with heavy metals, endangering aquatic life and human communities.

- **Manufacturing Processes:** Industrial manufacturing processes emit pollutants such as greenhouse gases (GHGs), volatile organic compounds (VOCs), and particulate matter, all of which impair air quality and contribute to climate change. Textile manufacturing, for example, contributes significantly to water pollution due to the use of dyes and chemicals.

- **Waste Generation:** The disposal of consumer products, particularly non-biodegradable materials such as plastics, presents considerable waste management difficulties. Landfills and incineration not only take up precious land but also emit dangerous pollutants into the environment. Ocean plastic pollution is a particularly serious problem, harming marine life and ecosystems.

Addressing these environmental consequences necessitates a move toward sustainable production and consumption habits. This involves encouraging the circular economy, in which items are built for durability, reuse, and recycling, reducing waste and resource extraction.

Social injustices

The global supply chains that underpin current consumer culture frequently contain exploitative labor practices and social inequalities. Vulnerable communities, especially in poor nations, bear these hidden expenses.

- **Labor Exploitation:** Many consumer items are manufactured in low-wage nations where workers endure harsh working conditions, insufficient pay, and a lack of labor rights. The fast fashion business, for example, relies on sweatshop labor to make low-cost apparel, which often comes at the expense of workers' health and well-being.

- **Child Labor:** The need for low-cost commodities can lead to the exploitation of children in areas including mining, agriculture, and manufacturing. Children who work in hazardous conditions are denied an education and face bodily and psychological harm.

- **Community Displacement:** Large-scale industrial operations, such as mining and infrastructure development, can uproot communities, robbing them of their land and livelihoods. Indigenous groups are more vulnerable to displacement and the loss of cultural heritage.

Addressing these social inequities entails campaigning for fair trade practices, assisting companies with ethical labor standards, and implementing legislation that safeguards workers' rights. Consumers may contribute by making informed decisions and demanding openness in supplier chains.

Health effects

The hidden costs of consumption also affect human health. Many products and habits connected with modern consumer society have negative health consequences:

- **Toxic Chemicals:** The use of hazardous chemicals in agriculture, manufacturing, and domestic products can cause respiratory problems, skin irritation, and chronic disorders. Pesticides employed in conventional farming, for example, can pollute food and water, endangering farmworkers and customers.

- Air and water pollution: Industrial and transportation-related pollution contributes to respiratory and cardiovascular disorders. Air pollution, mostly caused by car emissions and industrial operations, is a primary cause of premature death globally. Water pollution caused by agricultural runoff, industrial discharges, and poor waste management can result in waterborne infections and long-term health problems.

- Processed Foods: The widespread availability of processed and fast foods, which are frequently heavy in sugar, salt, and harmful fats, has led to the global rise in obesity, diabetes, and cardiovascular disease. These food habits are connected to a convenience-driven consumer culture that values quick, low-cost meals over nutritious value.

Addressing Hidden Health Costs.

Mitigating these hidden health costs necessitates collaborative efforts from several sectors:

- **Chemical Regulation:** Stricter rules on the use of toxic chemicals in agriculture and industries can help to decrease exposure to harmful compounds. Policies should encourage the use of safer alternatives while imposing rigorous limitations on dangerous chemicals.

- **Improving Air Quality:** Investing in cleaner energy sources and expanding public transit can help to minimize air pollution. Cleaner air requires policies that encourage the use of electric vehicles while discouraging the use of fossil fuels.

- Promoting Healthy Diets: Public health campaigns and educational initiatives can raise awareness of the benefits of a well-balanced, whole-food diet. Encouraging the consumption of fresh, locally sourced vegetables rather than manufactured meals can enhance dietary patterns and health consequences.

The Role of Consumers

Consumers have enormous potential to drive change by making educated decisions and demanding better practices from businesses and governments. Here are several approaches for consumers to help solve the hidden costs of consumption:

- **Sustainable Purchasing:** Choose products that are sustainably sourced and ethically made. Look for certifications such as Fair Trade, Organic, and Forest Stewardship Council to show responsible production practices.

- **Supporting Ethical Brands:** Choose firms that promote sustainability and social responsibility. Researching and selecting businesses with transparent supply chains and ethical labor standards can result in industry-wide reforms.

- **Waste Reduction:** Aim for a zero-waste lifestyle by reducing single-use products, recycling, and composting. Reusing and repurposing products can also help to lessen the environmental impact of garbage.

- **Advocacy and Activism:** Participate in advocacy campaigns to promote stricter environmental and social regulations. Support legislation that encourages sustainability and fair labor standards, and get involved in efforts that raise awareness about the hidden costs of consumerism.

The Path Forward

Understanding the state of our world, the concept of carbon footprints and ecological imprints, and the hidden costs of consumption are critical to creating a sustainable future. However, awareness is not enough. It must be combined with action, both individual and collective.

Innovation and Solutions

Innovative solutions across several sectors provide hope for lowering environmental effects and resolving consumption's hidden costs:

- **Circular Economy:** This concept seeks to eliminate waste while encouraging the continuous use of resources. The circular economy can reduce resource extraction and waste generation by designing goods that are durable, reusable, and recyclable.

- **Renewable Energy:** Advancements in solar, wind, and other renewable energy sources have the potential to drastically cut greenhouse gas emissions. Investing in renewable energy infrastructure and boosting energy efficiency are key measures toward achieving a low-carbon future.

- **Sustainable Agriculture:** Techniques like regenerative farming, agroforestry, and permaculture can improve soil health, boost biodiversity, and sequester carbon. Supporting local and organic farming can help to lessen the environmental impact of food production.

- **Green Technologies:** Biodegradable materials, electric vehicles, and smart grids are examples of green technology innovations that show promise for minimizing environmental impact. Supporting R&D in these areas is critical for long-term progress.

Collective Action

Individual activities are vital, but collaborative efforts are critical for large-scale change. Collaboration between governments, companies, and civil society can create systemic change. Key areas for collective action are:

- **Policy and regulation:** Governments have an important role in establishing and enforcing environmental standards. Policies that promote renewable energy, sustainable agriculture, and waste reduction can help to build a more sustainable environment.

- **Corporate Responsibility:** Businesses must be accountable for their environmental and social repercussions. Corporate sustainability measures, transparency reporting, and ethical supply chain management are crucial for lowering the hidden costs of consumption.

- **Community Engagement:** Grassroots movements and community-led initiatives can help propel local sustainability efforts. Community gardens, renewable energy cooperatives, and local recycling programs are examples of how collective action may have a real impact.

Education and Awareness. Education is an effective technique for instilling a culture of sustainability. Raising awareness of environmental challenges and the hidden costs of consumption can motivate people and communities to act. The key educational initiatives include:

- **Environmental Education:** Incorporating environmental education into school curricula helps foster a sense of sustainability in students. Teaching kids about sustainability, conservation, and the value of biodiversity can help mold future generations of environmentally concerned people.

- **Public Awareness efforts:** These efforts can inform and engage the general public. Using media, social networks, and community events to emphasize the effects of consumption and the benefits of sustainable behaviors can help to encourage behavior change.

Personal Action Plan

Individuals may make real efforts to decrease their impact on the environment and contribute to a more sustainable future. Here's a personalized action plan to get started:

1. Calculate Your Carbon Footprint: Use internet tools to assess your carbon footprint and identify places where you may cut emissions. This could include minimizing energy use, using public transportation, or following a plant-based diet.

2. Support Renewable Energy: Choose renewable energy sources or, if possible, install solar panels. Reduced reliance on fossil fuels is an important step in lowering your carbon impact.

3. Practice sustainable practices by reducing, reusing, and recycling whenever possible. Reduce single-use plastics, compost organic waste, and promote recycled products.

4. Choose Ethical Products: Research and buy from companies that value sustainability and fair labor standards. Fair Trade and Organic certifications might help you make an informed decision.

5. Advocate for Change: Get engaged in local environmental projects and support legislation that encourages sustainability. Join or support groups that promote environmental protection and social justice.

6. Educate Yourself and Others: Stay informed about environmental issues and share what you know with others. Participate in educational initiatives and encourage your community to use sustainable practices.

6. Educate Yourself and Others: Stay informed about environmental issues and share your knowledge with others. Participate in educational programs and encourage your community to adopt sustainable practices.

Our Shared Responsibility

Understanding our impact on the planet is the first step towards creating a sustainable future. The state of our planet, the concept of carbon footprints and ecological imprints, and the hidden costs of consumption highlight the urgent need for change. By taking individual and collective action, supporting innovative solutions, and advocating for responsible policies, we can build a green legacy for future generations.

Chapter 2
Redefining Sustainability

Sustainability has become a catchphrase in recent years, appearing frequently in conversations about environmental conservation, corporate accountability, and economic progress. However, the underlying nature of sustainability is more than just language; it is a comprehensive approach to living and interacting with the earth in a way that promotes long-term health and viability for both human and natural systems.

Define Sustainability

At its essence, sustainability refers to the ability to meet our current demands without jeopardizing future generations' ability to do the same. This philosophy is founded on three interrelated pillars: environmental health, economic viability, and social equality. A truly sustainable approach strives to balance these three pillars, knowing that neglecting any of them diminishes the others.

1. Environmental Health: This pillar focuses on safeguarding and preserving natural resources, ecosystems, and biodiversity. It entails lowering pollution, conserving water and energy, limiting waste, and protecting habitats to ensure that the Earth's ability to support life is preserved.

2. Economic viability: Economic sustainability entails developing robust systems capable of generating long-term prosperity. This includes encouraging innovation, maintaining fair trade practices, boosting local economies, and creating economic models that do not deplete natural resources.

3. Social Equity: The social side of sustainability highlights the importance of diverse and equitable communities. It addresses topics such as education, healthcare, and basic amenities, ensuring that all people have the opportunity to succeed without being discriminated against or treated unfairly.

Interconnected Challenges

The interconnectedness of these pillars necessitates holistic solutions. For example, encouraging renewable energy addresses environmental concerns while also providing economic advantages by creating jobs and lowering energy costs. Similarly, supplying clean water promotes public health (social fairness) and increases productivity (economic viability).

Sustainability in Practice

Implementing sustainability entails both individual and systemic changes. Individual choices such as decreasing energy use, recycling, and supporting ethical products help to ensure sustainability. Policies that encourage renewable energy, sustainable agriculture, and equitable economic development must be implemented on a systemic scale.

- **Renewable Energy:** Using renewable energy sources like solar and wind power lessens our dependency on fossil fuels and helps to combat climate change. This change also promotes economic sustainability by producing green jobs and lowering energy prices in the long run.

- **Sustainable Agriculture:** Practices including crop rotation, organic farming, and permaculture improve soil health, conserve water, and boost biodiversity. These measures not only safeguard the environment but also promote food security and rural economies.

- **Circular Economy:** The circular economy model, which we shall discuss in depth in the following section, focuses on resource efficiency and waste reduction. This strategy reduces environmental effects while fostering economic growth via innovation and new business opportunities.

- **Equitable Access:** Ensuring that all communities have access to clean water, education, healthcare, and economic opportunities is critical for social sustainability. Policies addressing poverty, injustice, and inequality are critical for fostering inclusive societies.

By combining these techniques, we may build resilient, adaptable systems that can support both current and future generations.

Principles of the Circular Economy
The old linear economic model, which employs a "take, make, dispose" strategy, has resulted in considerable environmental damage and resource depletion. In contrast, a circular economy seeks to redefine growth by emphasizing beneficial societal benefits. It requires gradually disconnecting economic activity from the use of scarce resources and eliminating waste from the system.

Core Principles of the Circular Economy

1. Eliminate Waste and Pollution: In a circular economy, items are created with the entire lifecycle in mind. This entails designing products that are durable, repairable, and recyclable, reducing waste and environmental impact from the start. Materials should be chosen based on their reusability or safe return to the environment.

2. Keep Products and Materials in Use: The circular economy relies heavily on extending the lives of products and materials. This includes tactics like maintenance, repair, refurbishing, and remanufacturing. By keeping items and materials in circulation, we limit the demand for new resources and waste.

3. Regenerate Natural Systems: A circular economy attempts to inflict less harm while simultaneously creating beneficial consequences for natural systems. This encompasses techniques that restore and improve ecosystems, such as regenerative agriculture, which promotes soil health and biodiversity.

Circular Economy in Action
- Product Design: Businesses are increasingly implementing design concepts that promote repair,

recycling, and resource efficiency. For example, modular cell phones enable customers to change individual components rather than the complete device. Similarly, clothing companies are producing outfits using recycled materials and providing take-back services for unwanted clothes.

- **Resource Recovery:** New technologies and business models are being developed to recover valuable materials from waste streams. For example, urban mining involves recovering metals from electronic garbage, which reduces the need for virgin mining and has a lesser environmental impact.

- **Sharing Economy:** The sharing economy, represented by platforms such as Airbnb and Uber, exhibits circular ideas by making the best use of existing assets. By sharing resources, we lessen the demand for new products and overall consumption.

- **Industrial Symbiosis:** This notion involves multiple industries working together to use each other's byproducts. For example, waste heat from a power plant can be utilized to heat nearby greenhouses, resulting in a symbiotic connection that benefits both parties while reducing waste.

Benefits of the Circular Economy

- **Environment:** By limiting resource exploitation and waste, the circular economy reduces environmental impact, conserves natural resources, and reduces greenhouse gas emissions.

- **Economic:** The circular economy promotes innovation, opens up new business opportunities, and generates jobs. It also improves resource security by decreasing reliance on finite resources.

- **Social:** Circular practices can strengthen communities by creating jobs and reducing environmental dangers. Additionally, emphasizing product durability and repairability can make items more affordable and accessible.

Challenges and Opportunities

To transition to a circular economy, various hurdles must be overcome, including technological barriers, legislative frameworks, and consumer behavior. However, the prospective rewards make it a worthy undertaking. Key measures to aid this transition are:

- **Policy Support:** Governments can encourage circular behaviors by implementing rules and regulations that promote recycling, resource efficiency, and sustainable product design.

- **Collaboration:** The creation of circular systems requires cross-sector collaboration. Businesses, governments, and civil society must collaborate to create and implement circular solutions.

- **Education and understanding:** Raising understanding of the benefits of a circular economy can increase customer demand for sustainable products and practices. Educational programs can also provide individuals with the necessary skills to engage in the circular economy, such as repair and recycling processes.

Chapter 3
Renewable Energy and the Way Forward.

As the globe grapples with serious issues such as climate change, depleting fossil fuel supplies, and environmental degradation, the transition to renewable energy sources has become a necessity rather than an option. Renewable energy sources, which are abundant, sustainable, and have a low environmental impact, provide a potential solution to our energy requirements. This chapter delves into the many renewable energy sources, advances in energy storage and efficiency, and the crucial role of policy and activism in accelerating the energy transition.

The Promise of Renewable Energy
Unlike fossil fuels, which are finite and contribute to greenhouse gas emissions, renewable energy sources are clean, limitless, and becoming more cost-competitive. Solar, wind, hydro, geothermal, and biomass are

the five most common forms of renewable energy.

Solar Energy

Solar energy is derived from the sun's radiation, which is plentiful and free. Photovoltaic cells turn sunlight directly into energy, whereas solar thermal systems heat fluids to produce steam and power turbines.

- **Photovoltaic Cells:** Advances in Photovoltaic cells technology have dramatically increased the efficiency and affordability of solar panels. Innovations like perovskite solar cells and bifacial panels, which catch sunlight from both sides, are pushing the limits of solar efficiency and cost-effectiveness.

- **Concentrated Solar Power:** Concentrated Solar Power systems employ mirrors or lenses to focus sunlight on a narrow area, resulting in high temperatures that generate steam to power turbines.Concentrated Solar Power can include thermal energy storage, enabling for power generation even when the sun isn't shining.

Wind Energy

Wind farms can be built onshore or offshore, and offshore wind farms benefit from stronger and more consistent winds.

Onshore wind farms are extremely simple to establish and maintain, making them a popular alternative in many nations. Onshore wind energy efficiency and capacity have increased due to advances in turbine design, such as taller towers and larger blades.

- **Offshore Wind:** Wind farms placed on bodies of water benefit from higher wind speeds and lower visual effects. Floating wind turbines are a new technology that enables the establishment of wind farms in deeper waters, where wind resources are more plentiful.

Hydropower

Hydropower is one of the oldest and most popular kinds of renewable energy.

- **Large Hydropower:** Large hydroelectric dams produce significant amounts of electricity and can supply dependable baseload power. However, they can have serious environmental and social consequences, including habitat loss and community dislocation.

- **Small Hydropower:** Small and micro hydropower systems have a low environmental impact and can power isolated or off-grid communities. Run-of-river systems, which do not require massive dams, are an eco-friendly option.

Geothermal Energy

Geothermal energy uses heat from the Earth's interior to produce power or give direct heating. It is a dependable and consistent source of energy, with the potential for extensive application in areas with high geothermal activity.

- **Geothermal Power Plants:** These plants create power by driving turbines with steam produced from hot water reservoirs located a few miles below the Earth's surface.

- **Direct Use Applications:** Geothermal heat can be applied directly to a variety of applications, including district heating, greenhouse heating, and industrial processes.

Biomass Energy

Biomass energy comes from organic resources like wood, agricultural waste, and animal waste. It can be used to generate power, heat buildings, and produce biofuels.

- **Bioenergy Power Plants:** These plants burn biomass to generate steam, which then powers turbines to generate electricity. Co-firing biomass and coal in existing power stations is a transitional method that can help reduce carbon emissions.

- **Biofuels:** Biofuels, such as ethanol and biodiesel, are derived from biomass and can be used to replace gasoline and diesel. Advanced biofuels derived from non-food crops and waste materials provide better sustainability and lower environmental impact.

The benefits of renewable energy
Renewable energy sources have many advantages, including:

- **Environmental Advantages:** Renewable energy reduces greenhouse gas emissions, air pollution, and water use when compared to fossil fuels. It also reduces the effects of climate change while preserving natural habitats.

- **Economic Benefits:** Renewable energy generates jobs, supports economic growth, and lowers energy costs over time. Renewable energy investments have the potential to drive technical innovation while also improving energy security.

- **Social Benefits:** Access to clean energy promotes public health, alleviates poverty, and raises quality of life. Renewable energy initiatives can help communities, particularly in distant and underdeveloped locations.

Innovation in Energy Storage and Efficiency

While renewable energy sources are critical to a sustainable future, their intermittent nature creates obstacles to constant and reliable energy delivery. Innovations in energy storage and efficiency are required to overcome these issues and facilitate a smooth transition to a renewable energy-based system.

Energy Storage Technologies

Energy storage devices store excess energy supplied by renewable sources for later use when demand is high or generation is low. Advances in energy storage are crucial for integrating renewables into the grid and maintaining a stable electricity supply.

- **Batteries:** The most extensively utilized energy storage technology is lithium-ion batteries, which have a high energy density and efficiency. Solid-state batteries, which use solid electrolytes rather than liquids, promise increased safety, longevity, and energy density.

- **Flow Batteries:** Flow batteries store energy in liquid electrolytes held in external tanks, allowing for a variable storage capacity. They are ideal for large-scale applications and have long cycle times.

- **Pumped Hydro Storage:** This technology employs surplus electricity to pump water uphill to a reservoir. When energy is required, the water is freed to run downhill via turbines, which generate electricity. Pumped hydro storage is the most established and commonly used type of large-scale energy storage.

- **Compressed Air Energy Storages:** These systems compress air and store it underground in caverns or tanks using excess electricity. When electricity is needed, the compressed air is released and heated to power turbines

Energy Efficiency Innovations

Improving energy efficiency is an important method for lowering energy demand and increasing the usage of renewable energy. Energy efficiency innovations have a wide range of applications, including buildings, transportation, and industry.

- **Building Efficiency:** Improvements in building design, materials, and technology can considerably cut energy use. Passive house standards, green building certifications, and smart building technologies all improve energy efficiency in heating, cooling, lighting, and appliances.

- **Transportation Efficiency:** The development of electric vehicles, fuel-efficient engines, and innovative public transit systems is transforming the transportation industry. Vehicle efficiency is improved through innovations such as lightweight materials, aerodynamic designs, and regenerative brakes.

- **Industrial Efficiency:** Industrial operations account for a large share of world energy consumption. Innovations such as combined heat and power systems, waste heat recovery, and energy-efficient manufacturing practices can help minimize industrial energy consumption and emissions.

Smart Grids & Demand Response

Smart grids use digital technologies to improve the electrical grid's dependability, efficiency, and flexibility. They allow for real-time monitoring and management of energy flows, simplifying the integration of renewable energy and increasing grid resilience.

- **Advanced Metering Infrastructure :** Smart meters give real-time data on energy consumption, allowing consumers to make more informed decisions and utilities to optimize grid operations.

- **Demand Response:** Demand response programs encourage users to reduce or shift their energy consumption during peak hours, thereby balancing supply and demand. This can be accomplished using automated systems, time-of-use pricing, and user participation.

- **System-scale Storage:** Integrating large-scale energy storage into smart grids can improve system stability, provide backup power, and make it easier to integrate renewable energy sources.

The Impact of Policy and Advocacy in Energy Transition

To overcome impediments and create systemic change, a transition to renewable energy requires supporting policies and vigorous lobbying. Governments, industry, and civil society must collaborate to establish a favorable climate for renewable energy development and deployment.

Effective policies can hasten the adoption of renewable energy by offering incentives, removing barriers, and establishing ambitious goals. Key policy instruments are:

- **Renewable Energy Targets:** Governments can establish binding targets for renewable energy generation, paving the way for future development. Examples include renewable energy directives from the European Union and renewable portfolio standards at the state level in the United States.

- Financial Incentives: Subsidies, tax credits, and grants can help lower the cost of renewable energy projects and make them more competitive with fossil fuels. Examples in the United States include the Investment Tax Credit and the Production Tax Credit.

- **Feed-in Tariffs:** FiTs ensure that renewable energy providers receive fixed payments for the power they create, giving long-term financial stability. This policy has been successful in nations such as Germany and Denmark.

- **Carbon Pricing:** Carbon pricing methods, such as carbon taxes and cap-and-trade systems, internalize the environmental costs of carbon emissions, making renewable energy more appealing than fossil fuels.

- **Regulatory Reforms:** To facilitate the expansion of renewable energy, permitting processes must be streamlined, grid infrastructure updated, and impediments to distribution and storage removed. Regulatory reforms should also address grid access, market structures, and utility business models to facilitate the transition to a decentralized and renewable-based energy system.

International Cooperation and Agreements

Global concerns necessitate global responses. International collaboration and agreements are critical in accelerating renewable energy production and combating climate change.

- **Paris Agreement:** Adopted in 2015, the Paris Agreement commits governments to keeping global warming well below 2 degrees Celsius above pre-industrial levels, with attempts to limit the increase to 1.5 degrees Celsius. This agreement establishes a framework for countries to improve their climate policies, particularly increasing renewable energy deployment.

- **International Renewable Energy Agency:** it promotes the broad and sustainable use of renewable energy around the world. It serves as a platform for international collaboration, knowledge sharing, and technical assistance in support of renewable energy transitions.

- The Clean Energy Ministerial is a global event that brings together energy ministers and stakeholders to advocate for policies and programs that advance clean energy technologies and solutions. It encourages collaboration on issues such as energy efficiency, smart grids, and renewable energy integration.

The Function of Advocacy in Civil Society

Advocacy and civil society play critical roles in accelerating the renewable energy transition. They increase awareness, shape legislation, and rally public support for sustainable energy programs.

- Environmental organizations include Greenpeace, the Sierra Club, and Friends of the Earth, which campaign for renewable energy policy, oppose fossil fuel projects, and promote environmental justice.

They use campaigns, litigation, and policy lobbying to effect change.

- **Grassroots Movements:** Local and community-based organizations are leading the renewable energy revolution. Community solar projects, local clean energy cooperatives, and grassroots efforts for renewable energy policies all empower communities and put pressure on change from the ground up.

- **Youth Activism:** Young people are increasingly taking the lead on climate action and renewable energy. Fridays for Future, led by Greta Thunberg, and the Sunrise Movement in the United States have raised global awareness of the severity of the climate catastrophe and the necessity for aggressive renewable energy policy.

- **Corporate Advocacy:** Businesses and industry associations can encourage renewable energy adoption by making sustainability commitments, purchasing renewable energy, and advocating for supporting regulations. Initiatives like the RE100 initiative, in which businesses commit to using 100% renewable electricity, highlight the private sector's participation in the energy transformation.

The intersection of Equity and Energy Transition

Ensuring that the renewable energy transition is just and equitable is critical to its success. Policies and initiatives must address concerns about energy access, affordability, and social justice.

- **Energy Access:** Approximately 770 million people globally still do not have access to power. Renewable energy, particularly localized solutions like solar household systems and mini-grids, has the potential to provide underprivileged communities with affordable and reliable energy, thereby enhancing their quality of life and economic possibilities.

- **Energy Affordability:** Policymakers must ensure that the expenses of shifting to renewable energy do not fall disproportionately on low-income households. Targeted subsidies, energy efficiency programs, and inclusive financing structures can all contribute to making renewable energy more accessible and affordable for everyone.

- **Just Transition:** The transition from fossil fuels to renewable energy will have a substantial impact on workers and communities that rely on traditional energy sources. A just transition framework aims to support these people and communities through retraining, economic diversification, and social protection measures.

Success Stories & Case Studies
Several countries and regions have achieved substantial progress toward renewable energy transitions, providing useful lessons and encouragement to others.

- **Germany:** The German energy transition is a comprehensive policy framework aiming at transforming the country's energy system into one that is low-carbon and nuclear-free. The country has invested heavily in renewable energy, energy efficiency, and grid upgrading. As a result, Germany has become a world leader in renewable energy deployment, particularly wind and solar power.

- **Denmark:** Denmark has had great success integrating renewable energy into its grid. The country generates roughly half of its electricity from wind power and has set lofty goals for becoming fossil fuel-free. Denmark's success is due to supporting policies, public engagement, and a significant emphasis on energy efficiency.

- **Costa Rica:** More than 98% of its electricity comes from renewable sources such as hydroelectric, wind, geothermal, and solar. The country's dedication to sustainability and environmental conservation has established it as a pioneer in renewable energy and climate action.

- **China:** As the world's top greenhouse gas emitter, China understands the significance of shifting to renewable energy. The country has emerged as the world's largest producer of solar panels and wind turbines, and it has made major expenditures in renewable energy deployment and innovation. China's renewable energy capacity continues to expand, aiding global efforts to tackle climate change.

The Way Forward: A Renewable Future.

The shift to a renewable energy future is not without challenges, but the benefits outweigh the drawbacks. We can build a sustainable, resilient, and egalitarian energy system by leveraging renewable energy sources, improving energy storage and efficiency, and enacting supportive legislation.

Key Actions for a Renewable Future.

- Continued investment in research and development is critical for expanding renewable energy technology, increasing efficiency, and lowering costs. Governments, corporations, and research institutions must work together to foster innovation and bring new ideas to market.

- **Increase Deployment:** Expanding the deployment of renewable energy projects is crucial to reaching climate targets and reducing dependency on fossil fuels. To expedite the adoption of renewable energy, supportive policies, financial incentives, and faster permitting processes are needed.

- **Improve Grid Infrastructure:** Modernizing and expanding grid infrastructure is required to accommodate greater amounts of renewable energy. This includes improving transmission and distribution networks, adopting smart grid technology, and creating interconnections to enable cross-border energy flows.

- **Encourage Energy Efficiency:** Improving energy efficiency in all industries reduces energy demand while complementing renewable energy installations. Policies and programs to encourage energy efficiency measures, such as building retrofits, efficient appliances, and industrial processes, are critical.

- Foster International Cooperation: Addressing the climate issue and promoting renewable energy require global collaboration. Countries must collaborate to share knowledge, technologies, and best practices, as well as to assist underdeveloped countries in making the transition to renewable energy.

- Engage and Empower Communities: Getting communities to participate and benefit from the renewable energy transition is critical to its success. This includes funding community energy projects, offering education and training, and addressing social and economic consequences through a just transition framework.

Chapter 4
Creating a Sustainable Future: Transforming Our Food Systems

The Journey from Farm to Fork

Food's journey from farm to fork is a complicated and linked process that includes several stages such as production, processing, distribution, consumption, and disposal. Understanding this path is critical for identifying ways to improve our food systems' sustainability, resilience, and equity.

Agricultural production

Agricultural production is the initial phase in the food supply chain. Sustainable farming practices are critical for protecting natural resources, improving soil health, and minimizing environmental damage.

- **Sustainable Agriculture:** Practices including crop rotation, agroforestry, and integrated pest control increase biodiversity, improve soil fertility, and reduce the need for chemical inputs. Organic farming, which does not use synthetic fertilizers or pesticides, is another approach that promotes environmental health and food safety.

- **Regenerative Agriculture:** Regenerative agriculture goes beyond sustainability, aiming to restore and improve ecosystems. No-till farming, cover cropping, and holistic grazing techniques all contribute to soil carbon sequestration, increased water retention, and biodiversity.

- **Precision Agriculture:** Using technology to optimize agricultural techniques can boost efficiency and cut waste. Precision agriculture employs satellite, drone, and sensor data to monitor crop health, soil conditions, and weather patterns, allowing farmers to make more informed irrigation, fertilization, and pest management decisions.

Processing and Packaging

Once gathered, food is processed to convert raw materials into edible items. Cleaning, sorting, cooking, preserving, and packaging are all examples of processing operations. Innovations at this stage can improve food safety, minimize waste, and increase sustainability.

- **Minimal Processing:** Minimally processed foods retain more nutrients and use fewer resources than fully processed items. Encouraging the intake of fresh, whole foods can help to lessen our diet's environmental imprint.

- **Sustainable Packaging:** Packaging has a tremendous impact on waste and pollution. Innovations in biodegradable, compostable, and recyclable packaging materials can help to lessen environmental impact. Companies are also looking at edible packaging and reusable containers as alternatives to single-use plastics.

- **Food Preservation:** Techniques like canning, freezing, and drying increase the shelf life of food, reducing spoilage and waste.

Advances in preservation technology, such as high-pressure processing and vacuum sealing, can enhance food safety and quality while using less energy.

Distribution & Logistics

Food distribution is getting products from farms and processing plants to stores, restaurants, and customers. Efficient and sustainable logistics are critical for reducing food miles, lowering emissions, and ensuring that food reaches people who need it.

- **Local Food Systems:** Promoting local food systems can shorten the supply chain, lower transportation emissions, boost local economies, and improve food security. Farmers' markets, community-supported agriculture programs, and farm-to-table initiatives allow consumers to interact directly with producers.

- **Cold Chain Management:** Proper temperature control throughout the supply chain is crucial to protecting the quality and safety of perishable items. Refrigeration technology innovations, such as solar-powered coolers and phase-change materials, have the potential to increase cold chain efficiency and sustainability.

- **Smart Logistics:** By leveraging data analytics, artificial intelligence, and blockchain technology, businesses can optimize logistics, decrease waste, and increase transparency. These technologies allow for better inventory management, route optimization, and traceability, ensuring that food is delivered efficiently and safely.

Consumption and Dietary Choices
Consumer decisions and eating habits have a significant impact on the sustainability of our food systems. Promoting healthy and sustainable diets can help to reduce environmental impact, promote public health, and encourage ethical food production.

- **Plant-Based Diets:** Plant-based diets, which include fruits, vegetables, legumes, nuts, and whole grains, have a lesser environmental effect than animal-based diets. Reducing meat and dairy intake can help to reduce greenhouse gas emissions, land use, and water use.

- **Mindful Eating:** Promoting mindful eating habits can assist in preventing overconsumption and food waste. This involves keeping track of portion sizes, savoring food, and avoiding distractions when eating.

- **Ethical Consumption:** By purchasing organic, fair trade, or certified items that meet sustainability standards such as the Rainforest Alliance or the Marine Stewardship Council, consumers may help to ensure sustainable and ethical food production. Supporting businesses that promote animal care, fair labor standards, and environmental sustainability is also critical.

Waste and Disposal

Waste and disposal mark the end of the food journey. To minimize the environmental impact of our food systems, we must reduce food waste and develop sustainable waste management procedures.

- **Food Recovery and Redistribution:** Excess food from

farms, shops, and restaurants can be redistributed to food banks, shelters, and community kitchens to combat food hunger and waste. These efforts rely heavily on initiatives like gleaning programs and food rescue organizations.

- **Composting and Anaerobic** Digestion: Organic waste can be composted to make nutrient-rich soil amendments or anaerobically digested to produce biogas and compost. These measures redirect food waste from landfills, lowering methane emissions.

- **Circular Food Systems:** Implementing a circular economy approach to food systems entails reducing waste, reusing products and materials, and renewing natural systems. This can involve converting food waste into animal feed, developing bio-based goods, and incorporating agroecological methods.

Urban Farming and Community Gardens

Urban farming and community gardens are innovative approaches that bring food production to cities and communities. These programs improve food security, promote environmental sustainability, and foster stronger community bonds.

Benefits of Urban Farming

Urban gardening has various advantages, including:

- **Food Security:** Growing food in cities improves access to fresh, nutritious produce, especially in food deserts where healthy eating options are scarce. Urban farming can enhance household food sources while reducing dependency on distant supply lines.

-**Environmental Sustainability:** Urban farms and gardens help to green up cities, improve air quality, and reduce urban heat islands. They also enhance biodiversity, soil health, and water management by using measures like rainwater gathering and composting.

- **Community Building:** Urban farming promotes social relationships, community involvement, and a sense of belonging and pride. Community gardens offer places for people to connect, learn, and interact, thus increasing social cohesion and resilience.

- **Education and Skill Development:** Urban farming programs provide opportunity for both children and adults to learn about agriculture, nutrition, and sustainability. They can also offer career training and entrepreneurial opportunities in the rapidly expanding field of urban agriculture.

Types of Urban Farming

Urban farming comprises a wide range of approaches and systems, each tailored to particular urban environments and scales.

- **Community Gardens:** These communal places are used by community members to raise vegetables, fruits, and flowers. Community gardens can be found on vacant lots, roofs, or public locations, and they are frequently administered by local organizations or volunteer groups.

- **Rooftop Gardens:** Gardening on rooftops makes better use of underutilized urban spaces. Rooftop gardens range from small container gardens to large green roofs that promote food production, stormwater management, and energy efficiency.

- Vertical farming entails growing crops in stacked layers or vertically inclined surfaces, typically utilizing hydroponic or aeroponic systems. This technology maximizes space efficiency and can be used in buildings, shipping containers, and purpose-built structures.

- **Hydroponics and Aquaponics:** Hydroponic systems use nutrient-rich water to grow plants without the need for soil. Aquaponics combines hydroponics and aquaculture to form a symbiotic system in which fish waste feeds nutrients to plants while plants filter the water for fish.

- **Urban Greenhouses:** Greenhouses in urban locations lengthen the growing season and protect crops from harsh weather. They may be utilized for a wide range of crops and are compatible with renewable energy and water recycling systems.

Case Studies & Success Stories

Several towns and communities throughout the world have successfully undertaken urban farming programs, illustrating the potential of this approach.

- **Detroit, USA:** The city has embraced urban agriculture as a rehabilitation and food security plan. The city is home to several community gardens, urban farms, and organizations like Keep Growing Detroit, which promotes local food production and access.

- **Havana, Cuba**: Faced with food scarcity and limited resources, Havana created a thriving urban agriculture community.

Organopónicos, or urban organic farms, generate a large amount of the city's fresh produce, helping to ensure food security and sustainable growth.

- **Singapore:** As a densely populated city-state with limited land, Singapore has invested in novel urban farming methods. Sky Greens, a vertical farming firm, and Citizen Farm, a community-focused urban farm, are paving the way for sustainable food production.

Reducing Food Waste: Strategies and Solutions

Food waste is a serious problem with substantial environmental, economic, and social consequences. Globally, one-third of all food produced is wasted, which contributes to greenhouse gas emissions, resource depletion, and food poverty. Managing food waste necessitates a multifaceted approach that includes prevention, recovery, and recycling.

Prevention Strategies

The most efficient strategy to reduce food waste is to prevent it from occurring in the first place. This includes interventions at numerous points along the food supply chain.

- **Improved Harvesting and Storage:** In underdeveloped nations, inadequate harvesting procedures and storage facilities lead to severe food waste. Investing in enhanced infrastructure, like as cold storage and transportation, can help to reduce post-harvest losses.

- **Supply Chain Optimization:** Data analytics may help businesses better predict demand, optimize inventory management, and reduce overproduction. Collaboration among manufacturers, merchants, and consumers can also assist balance supply and demand while reducing waste.

- **Consumer Education:** Raising awareness about the causes and consequences of food waste can help consumers make more sustainable decisions. Meal planning, adequate food storage, and creative leftover use are all practices that can be promoted through educational programs.

Food Recovery and Redistribution
Surplus food that is safe to consume can be distributed to those in need, minimizing waste and solving food insecurity. Initiatives for food recovery and redistribution are critical components of this strategy.

Food banks and pantries

Food banks and pantries gather extra food from farms, retailers, and manufacturers and distribute it to individuals and families experiencing food insecurity. These groups rely on donations and partnerships with corporations and volunteers to function efficiently.

- Examples: Feeding America is a national network of food banks in the United States that distributes billions of meals each year. The Trussell Trust runs a network of food banks around the UK, offering emergency food and help to people in need.

Food Rescue Organizations

Food rescue organizations collect extra food from a variety of sources, including restaurants, supermarkets, and gatherings, and distribute it to those in need. These groups frequently operate with the assistance of volunteers and use technology to handle logistics.

- For example, New York City's City Harvest recovers millions of pounds of food each year and distributes it to community initiatives. In Australia, OzHarvest collects extra food from commercial sources and gives it to charities.

Learning Programs

Gleaning is removing excess crops from fields that would otherwise go to waste. Volunteers frequently collaborate with farmers to collect surplus produce and deliver it to food banks, shelters, and community kitchens.

- For example, the Society of St. Andrew arranges gleaning events around the United States, recovering millions of pounds of vegetables each year. Feedback Global in the United Kingdom performs gleaning efforts to collect surplus fruits and vegetables for redistribution.

Corporate Responsibility

Businesses can implement procedures and policies to reduce food waste and promote sustainability throughout their operations.

- Supermarkets and food sellers can reduce waste by improving inventory management, donating excess food, and offering discounts on products that are about to expire. Initiatives like "ugly produce" campaigns, which offer visually flawed but completely edible fruits and vegetables, can help to minimize waste.
- Restaurants and food service providers can reduce waste by implementing measures including portion management, trash tracking, and food donation. Programs such as

the National Restaurant Association's "Waste Not, Want Not" initiative give information and guidance for decreasing food waste in the business.

Individual actions
Individuals can help reduce food waste by being attentive to their intake and adopting sustainable behaviors.

Meal Planning and Shopping
Planning meals ahead of time and making shopping lists will help you avoid overspending and prevent food wastage.

- **Tip:** Make a list of necessary products before going shopping, buy only what you need, and prevent impulse purchases. Plan meals around perishable foods and use leftovers creatively.

Proper Storage

Proper food storage can improve the shelf life of items while reducing spoilage.

- **Tips:** Refrigerate or freeze perishable items, use airtight containers for dry goods and follow produce storage standards. Understanding the differences between "use by," "best before," and "sell by" dates can also aid in waste reduction.

Creative Cooking

Using leftovers and creatively repurposing items can help to reduce waste and make better use of available food.

- Tips: Make broth from vegetable leftovers, utilize leftover grains in salads or stir-fries, and turn overripe fruits into smoothies or baked goodies. Websites and applications that provide recipes based on readily available items can be useful sources of inspiration.

Composting At Home

Home composting may divert food scraps from landfills while producing beneficial compost for gardening.

- Tip: Create a compost bin or pile in your backyard, or use a worm composting system indoors. Many cities provide composting workshops and tools to help residents get started.

The Path Forward: A Sustainable Food System

Transforming our food systems to be more sustainable, resilient, and egalitarian is critical for addressing global issues like climate change, food insecurity, and biodiversity loss. By reconsidering how we produce, distribute, consume, and dispose of food, we may establish a food system that benefits both people and the environment.

Key Activities for a Sustainable Food System

- Promote Sustainable Agriculture: Help farmers use sustainable and regenerative techniques that improve soil health, reduce chemical inputs, and increase biodiversity. Provide education, incentives, and infrastructure to help with these transitions.

- **Reduce Food Waste:** Implement measures to prevent, recover, and recycle food waste at all stages of the supply chain. Encourage stakeholders to collaborate, create public awareness, and lobby for supporting policies.

- Support Urban Farming: Invest in urban agriculture projects that improve food security, promote environmental sustainability, and create community ties. Offer resources, instruction, and infrastructure to help urban farmers and gardeners.

- Encourage Sustainable Consumption: Encourage healthy and sustainable eating habits through education, public campaigns, and incentives. Encourage customers to limit their meat and dairy consumption, promote ethical food production, and reduce food waste.

- Improve Food Recovery and Redistribution: Expand food recovery and redistribution networks to guarantee that excess food reaches those in need. Fund food banks, food rescue organizations, and gleaning initiatives through donations, partnerships, and volunteer work.

- Promote Innovation and Technology: Encourage technology advancements that improve efficiency, minimize waste, and increase sustainability throughout the food supply chain. Support research and development in fields like precision agriculture, smart logistics, and waste management.

Chapter 5
Revolutionizing Transportation

Transportation is an important part of modern life since it allows people and things to travel throughout the world. However, it also has a significant impact on greenhouse gas emissions, air pollution, and urban congestion. As the globe grapples with the pressing need to address climate change and establish sustainable urban settings, transforming transportation infrastructure has emerged as a major priority. This chapter delves into the revolutionary possibilities of electric automobiles, public transportation improvements, and future mobility alternatives such as hyperloop and high-speed rail.

Electric Vehicles and Beyond

Electric vehicles represent a significant transition in transportation, providing a cleaner and more sustainable alternative to traditional internal combustion engine vehicles. Advances in battery technology, regulatory assistance, and increased consumer awareness of environmental issues are all driving the switch to electric vehicles.

The Rise of Electric Vehicles.
In recent years, electric vehicle sales have increased exponentially, reaching record highs around the world. Several causes contribute to this surge:

- **Technological Advancements:** Advances in battery technology have substantially boosted the range and performance of electric vehicles. Lithium-ion batteries, which are lighter and more efficient, have become the industry standard, and research into solid-state batteries promises even greater improvements in energy density and safety.

- **Government Policies and Incentives:** Numerous countries have created policies and incentives to encourage the use of EVs. These include tax breaks, rebates, and subsidies for electric vehicle purchases, as well as investments in charging infrastructure. Regulations such as pollution regulations and zero-emission car mandates help to push the change.

- **Environmental Awareness:** As people become increasingly aware of climate change and air pollution, they are looking for more environmentally friendly modes of transportation. EVs have zero tailpipe emissions, which reduces air pollutants that contribute to health issues and environmental deterioration.

Challenges and Opportunities

While the use of electric vehicles is increasing, numerous hurdles must be overcome to achieve widespread transition:

- **Charging Infrastructure:** The broad use of EVs requires a robust and accessible charging network. There are plans to enhance charging infrastructure, including fast-charging stations along roads and in urban areas. Wireless charging and vehicle-to-grid technology are also promising possibilities.

- **Battery Production and Recycling:** Lithium-ion battery production requires the extraction of raw materials such as lithium, cobalt, and nickel, which presents environmental and ethical considerations. Developing sustainable supply chains and optimizing battery recycling methods are critical for reducing the environmental effects of EVs.

- **Cost and Affordability:** Although the cost of EVs has reduced dramatically, they remain more expensive than conventional vehicles. Continued developments in battery technology and economies of scale are expected to drive down costs, making electric vehicles more affordable to a wider variety of consumers.

Beyond Electric Vehicles: The Next Frontier.

The electrification of transportation expands beyond passenger automobiles and into other modes of transportation, creating new prospects for sustainability.

- **Electric buses and trucks:** By electrifying public transit and freight vehicles, urban emissions can be greatly reduced. Electric buses are now in use in many cities, offering cleaner and quieter alternatives to diesel buses. Similarly, electric trucks are being developed for both short- and long-haul transportation, with businesses such as Tesla and Rivian leading the way.

- **Electric Aviation:** The aviation industry, a major contributor to greenhouse gas emissions, is looking into electrification as a way to minimize its carbon impact. Electric aircraft, such as the Aviation Alice and Vertical Aerospace VA-X4, are in various phases of development and testing. While electric aviation confronts technical problems, especially for long-haul flights, it shows promise for short-haul and regional travel.

- **Electric Marine Vessels:** The maritime industry is also electrifying, with electric ferries and cargo ships under development to minimize emissions from maritime transportation. Companies such as Norsepower and Yara Birkeland are pioneering the usage of electric and hybrid vessels, which employ renewable energy sources like wind and solar power to propel ships.

Public Transportation Innovations
Public transportation systems serve as the foundation of urban mobility, providing millions of people with efficient and inexpensive transportation options. Innovations in public transportation can improve the sustainability, accessibility, and convenience of these systems, lowering reliance on private automobiles and alleviating urban congestion.

Electrifying Public Transit
Electrifying public transportation systems, such as buses, trams, and trains, is an important technique for

lowering urban emissions and improving air quality.

- **Electric Buses:** Many cities are switching to electric buses to replace diesel fleets. Electric buses provide various advantages, including zero tailpipe emissions, lower operating costs, and a quieter operation. Cities like Shenzhen, China, have already completed the full electrification of their bus fleets, illustrating the feasibility and benefits of this move.

- Electrified light rail and tram systems offer efficient and sustainable urban transportation. These systems are powered by renewable electricity, which reduces their carbon footprint. Modern tram systems, such as those in Melbourne, Australia, and Portland, USA, blend perfectly into metropolitan landscapes, offering convenient and dependable transportation options.

Smart Public Transportation Systems

Smart technologies are revolutionizing public transportation, making it more efficient, user-friendly, and responsive to passenger demands.

- **Real-Time Information and Connectivity:** Providing real-time information about transit schedules, delays, and routes improves the passenger experience and increases the usage of public transportation. Mobile apps and digital displays at transit stations provide up-to-date information, allowing passengers to plan their trips more efficiently.

- Integrated Ticketing and Payment Systems: Combining ticketing and payment systems across several forms of transportation improves the user experience and encourages seamless travel. Contactless payment methods, such as mobile wallets and smart cards, enable travelers to pay for their journeys swiftly and efficiently.

- **Demand-Responsive transport:** These systems use technology to match transport supply with passenger demand in real-time. This strategy improves route planning and shortens wait times, resulting in a more flexible and efficient service. Examples include ride-hailing services, microtransit solutions, and on-demand shuttle services.

Sustainable Urban Mobility

Sustainable urban mobility entails planning cities and transportation systems that reduce dependency on private vehicles, encourage active travel, and create livable urban settings.

Complete Streets: The notion of "complete streets" entails creating streets that can handle all users, including walkers, bicycles, public transportation riders, and vehicles. This approach promotes safety, accessibility, and sustainability, resulting in lively and inclusive urban environments.

- Bike-Sharing and Micro-Mobility: Bike-sharing programs and micro-mobility solutions, such as e-scooters and e-bikes, offer practical and environmentally friendly alternatives to private vehicle use. These services complement public transportation by providing last-mile connectivity while reducing traffic congestion.

- Transit-Oriented Development: This is the planning and development of metropolitan regions around public transportation hubs. This strategy promotes higher-density, mixed-use development, reducing the demand for automobile traffic and encouraging sustainable land use.

The Future of Mobility: Hyperloops and High-Speed Rail

Future transportation technologies, such as hyperloop and high-speed rail, have the potential to transform long-distance travel by providing quick, efficient, and sustainable alternatives to traditional forms of transportation.

Hyperloops: Next-Generation Transportation

Elon Musk first proposed hyperloop technology in 2013, envisioning high-speed transport using vacuum tubes that would drastically reduce travel times and environmental impact.

How Hyperloops Work:

Hyperloops are made up of pods or capsules that move through low-pressure tunnels, decreasing air resistance and enabling high-speed transport. Magnetic levitation (maglev) technology is frequently utilized to lift and propel the pods, which reduces friction and energy consumption.

- **Speed and Efficiency:** Hyperloops can reach speeds of up to 760 mph (1,220 km/h), outpacing typical high-speed trains and competing with short-haul airplanes. Reduced friction and air resistance in vacuum tubes help to improve energy efficiency and cut operating expenses.

- **Environmental Impact:** Hyperloop technology has the potential to be more environmentally friendly than air and road travel. They can be fueled by renewable energy sources like solar and wind, resulting in low greenhouse gas emissions. The enclosed tube design also lowers noise pollution.

Challenges and Opportunities

While hyper loops provide great opportunities, numerous problems must be addressed to make them a reality:

- **Infrastructure and Cost:** Developing hyperloop infrastructure, such as vacuum tubes and stations, involves significant investment and engineering knowledge. The high initial expenditures and regulatory constraints provide significant barriers to wider deployment.

- **Safety and Comfort:** Keeping passengers safe and comfortable at high speeds is crucial. To acquire widespread adoption, hyperloop systems must handle challenges such as emergency evacuations, pressure variations, and ride smoothness.

- Regulation and Standardization: Establishing uniform standards and guidelines for hyperloop technology is critical to assuring safety, interoperability, and public trust. International engagement and alliances with governments, industry, and academia will be critical for resolving regulatory hurdles.

High-speed rail connects cities and regions.

High-speed rail systems provide a tried-and-true option for sustainable long-distance transport, connecting cities and regions with quick, efficient, and dependable service.

Advantages of High-speed Rail
- Speed and Convenience: High-speed trains can reach speeds of 186 mph (300 km/h) or higher, drastically reducing travel time between large cities. HSR stations are frequently positioned in city centers, offering easy access and minimizing the need for airport transfers.

- **Environmental Advantages:** Compared to vehicles and airlines, HSR systems emit fewer greenhouse gas emissions per passenger kilometer. Electrified HSR lines powered by renewable energy sources improve their environmental credentials.

- **Economic and social impact:** High-speed rail can boost economic growth, create jobs, and improve regional connectivity. It encourages tourism, business travel, and cultural interchange, so contributing to the social and economic vibrancy of connected regions.

Innovation and Future Developments in High-Speed Rail

The future of high-speed rail is being determined by ongoing developments aimed at increasing efficiency, sustainability, and passenger experience.

- Maglev Technology: Magnetic levitation (maglev) trains employ powerful magnets to lift and push the train, decreasing friction and allowing for faster speeds. Japan's SCMaglev is a prime example, with test runs reaching speeds of more than 600 km/h (373 mph). Maglev technology promises significantly faster journeys and smoother experiences.

- Energy Efficiency: High-speed train operators are increasingly concerned with energy efficiency and sustainability. Renewable energy sources, regenerative braking systems, and energy-efficient train designs can help to lessen high-speed rail's environmental impact.

- **Passenger Experience:** Improving passenger comfort, connection, and services is critical for drawing more travelers to high-speed rail. Innovative features like spacious seating, onboard Wi-Fi, and seamless ticketing systems all help to improve the travel experience.

The Impact of Policy and Advocacy on Transportation Transformation

Sustainable transportation systems necessitate collaborative efforts from governments, industry leaders, and the public. Policy and lobbying are critical in accelerating the transition to more sustainable and efficient transportation alternatives.

Government Policy and Incentives
Governments can create regulations and incentives to encourage the use of electric automobiles, public transportation technologies, and high-speed rail networks.

- **Subsidies and Tax Credits:** Offering financial incentives, such as subsidies and tax credits, can encourage the purchase of electric vehicles and the construction of charging stations. Incentives can also help with the expansion of high-speed rail and public transportation systems.

- **Regulations and regulations:** Setting strict emissions regulations and fuel efficiency criteria might encourage the car industry to adopt cleaner technologies. Creating standardized laws for emerging transportation technology like hyperloops and maglev trains promotes safety and compatibility.

- **Infrastructure Investment:** Governments can fund the creation and maintenance of transportation infrastructure such as electric vehicle charging stations, high-speed rail networks, and public transit systems. Public-private collaborations can also provide additional resources and expertise.

Public Awareness and Advocacy

Raising public awareness and creating advocacy is critical to gaining support for sustainable transportation measures.

- **Education Campaigns:** Public education campaigns can help citizens learn about the advantages of electric automobiles, public transportation, and high-speed rail. Highlighting the environmental, economic, and social benefits of sustainable transportation can motivate behavior change.

- **Community Engagement:** Involving communities in the planning and implementation of transportation projects ensures that solutions are tailored to local requirements and preferences. Public discussions, workshops, and participatory planning processes can help to increase community buy-in and support.

- **Advocacy Groups:** Advocacy groups and non-governmental organizations can play an important role in advancing sustainable transportation policies. These organizations can lobby for policy changes, undertake research, and organize grassroots campaigns to motivate action on a local, national, and international scale. The transformation of transportation systems is a critical component in creating a sustainable and resilient future. We can reduce emissions, alleviate urban congestion, and make cities more habitable by adopting electric vehicles, public transportation improvements, and future mobility alternatives such as hyperloop and high-speed rail.

The route to sustainable transportation necessitates a comprehensive approach that includes technology developments, supportive policies, public awareness, and community engagement. Collaboration between governments, industry leaders, researchers, and individuals is critical to driving this transformation

Chapter 6
Water for Life: Responding to the Global Water Crisis

Water is the foundation of life, and an essential resource for human survival, agricultural productivity, and industrial operations. However, the world is experiencing a growing water crisis, marked by scarcity, pollution, and unequal access. This chapter investigates the dimensions of the worldwide water crisis, looks at innovative water-saving technology, and considers policies for sustainable water use.

Global Water Crisis

The global water issue takes many forms, affecting millions of people and ecosystems globally. Understanding the scope and drivers of the situation is critical to developing successful remedies.

Water scarcity

Water scarcity arises when water demand exceeds supply. It is a major issue in many areas because of population increase, climate change, and ineffective water management.

- Physical water shortage arises when a region's natural water supplies are insufficient to meet demand. It is common in arid and semi-arid locations, including the Middle East, North Africa, and parts of South Asia. Countries such as Saudi Arabia and Yemen have acute physical water scarcity and rely largely on desalination and water imports.

- Economic water shortage is caused by a lack of investment in water extraction and distribution infrastructure or technologies. It frequently affects rural communities and underdeveloped countries, where financial and institutional hurdles impede adequate access to safe water. Sub-Saharan Africa is a prime example, with many populations relying on remote or unreliable water supplies.

Water Pollution

Water pollution exacerbates the water issue by contaminating freshwater sources, rendering them unfit for human consumption and harming ecosystems.

- **Chemical Contaminants:** Pollutants such as heavy metals, pesticides, and medicines are introduced into water bodies by industrial activity, agricultural runoff, and poor domestic chemical disposal. These toxins can cause health problems, damage aquatic life, and leave water unfit for drinking or irrigation.

- **Microbial Contaminants:** Microbial pollution from human and animal waste is a substantial health danger, especially in places with inadequate sanitation facilities. Waterborne diseases including cholera, dysentery, and typhoid fever can be caused by pathogens such as bacteria, viruses, or parasites.

- **Plastic Pollution:** Plastic waste is an increasing issue, with microplastics penetrating aquatic bodies and entering the food chain. These microscopic plastic particles can affect marine life and may have an impact on human health if consumed in seafood.

Unfair Access to Water

Access to clean and safe water is a basic human right, but billions of people do not have consistent access to this crucial resource.

- Urban-Rural Disparities: Urban areas typically have better access to water infrastructure than rural locations. However, increased urbanization can put a strain on existing systems, resulting in shortages and unpredictable supplies. Rural areas generally rely on untreated surface water or distant sources, making them more susceptible to water-related health risks.

- **Gender Inequities:** In many regions of the world, women and girls are primarily responsible for gathering water. This work frequently requires long, exhausting journeys, which limit their possibilities for education and economic activity. Ensuring fair access to water can empower women while also promoting broader social and economic growth.

- **Conflict and Displacement:** Water scarcity and pollution can escalate conflicts and cause displacement. Competition for scarce water supplies can create tensions between communities and countries. Furthermore, natural disasters and climate change-related events might relocate people, affecting access to clean water.

Innovative Water-saving Technologies

Addressing the global water crisis requires new solutions that improve water efficiency, decrease waste, and promote long-term use. Technological improvements provide exciting opportunities to save and manage water resources.

Advanced Irrigation Systems

Agriculture is the world's largest consumer of freshwater, accounting for over 70% of total water extraction. Improving irrigation efficiency is critical for lowering water use in agriculture.

- Drip irrigation feeds water directly to plant roots via a network of tubes and emitters, reducing evaporation and runoff. This approach has the potential to save up to 60% more water than typical flood irrigation. It is especially useful in dry climates where water conservation is critical.

- **Sprinkler Irrigation:** Sprinkler systems use overhead spray to simulate natural rainfall. While less effective than drip irrigation, contemporary sprinkler technology can save water through precise application and timing.

- **Smart Irrigation Systems:** By combining sensors, weather data, and automation, smart irrigation systems tailor water distribution to real-time conditions and crop needs. These devices have the potential to greatly improve water-use efficiency and reduce waste.

Water Recycling and Reuse

Recycling and reusing water can reduce the demand for fresh water and provide a long-term supply for a variety of uses.

- **Grey Water Recycling:** Grey Water from domestic activities like bathing, laundry, and dishwashing can be treated and reused for non-potable applications like irrigation and toilet flushing. This minimizes the demand for freshwater and the volume of effluent that needs to be treated.

- **Industrial Water Reuse:** Industries can employ water recycling processes to treat and reuse wastewater generated during operations. This reduces freshwater use and pollutant emissions into the environment. Industrial water treatment technologies include membrane filtration, reverse osmosis, and advanced oxidation.

- **Municipal Water Reuse**: Municipalities can purify wastewater to a high quality before reusing it for irrigation, landscape watering, and even drinking water. Examples include Orange County, California's Groundwater Replenishment System, which purifies wastewater to generate high-quality drinking water.

Desalination Technologies

Desalination is a solution for locations with limited freshwater resources since it converts seawater or brackish water into potable water.

- **Reverse Osmosis:** Reverse osmosis is the most commonly used desalination technology, which uses semi-permeable membranes to remove salts and contaminants from water. Advances in membrane technology and energy recovery have increased the efficiency and cost-effectiveness of RO desalination facilities.

- **Thermal desalination:** Heat is used to evaporate and condense water, separating it from salts. Examples include multi-stage flash distillation and multi-effect distillation. These technologies require a lot of energy, but they work well in areas with plenty of it.

- **Innovative Approaches:** Emerging desalination methods seek to improve efficiency while reducing environmental effects. Electro Dialysis, forward osmosis, and solar-powered desalination are all promising options that could change

Water Pricing and Economic Instruments

Economic measures, such as water pricing and subsidies, can encourage efficient water usage and promote long-term management.

- **Water Pricing:** Using pricing systems that reflect the full cost of water, including extraction, treatment, and delivery, promotes conservation and efficiency. Tiered pricing systems, in which higher consumption is paid at a higher rate, might encourage water conservation.

- **Subsidies and Incentives:** Offering financial incentives for water-saving devices and practices can encourage both individuals and businesses to invest in effective solutions. These incentives can help with the implementation of drip irrigation, greywater recycling, and other water-efficient methods.

- **Tradable Water Rights:** By creating markets for tradable water rights, water can be reallocated to areas where it is most needed and appreciated. This market-based strategy can improve water use efficiency and give flexibility in managing water resources during periods of scarcity.

Legislation and Governance

Strong legislative frameworks and strong governance structures are essential for ensuring sustainable water management and conserving water resources.

- Comprehensive water laws and regulations establish the legal framework for water usage, allocation, and protection. These laws should identify water rights, set quality standards, and provide methods for enforcement and dispute resolution.

- **Institutional Capacity:** Improving the capacity of water management institutions is critical to effective governance. This involves employee training, infrastructural improvements, and agency coordination. Building institutional capacity ensures that policies are carried out efficiently and water resources are managed sustainably.

- **Community Involvement:** Including local communities in water management decisions ensures that policies reflect the needs and interests of people who are most affected. Participatory initiatives, such as community-based water management committees, can improve accountability and promote environmentally friendly practices.

International Cooperation
Water resources frequently cross political lines, demanding international cooperation for efficient management and dispute resolution.

Countries that share transboundary water resources must work together to ensure long-term water availability and peace.

- **Transboundary Water Agreements:** Creating and implementing transboundary water agreements can assist manage shared water resources equitably and sustainably. These agreements should specify water allocation, quality criteria, and conflict resolution procedures. Successful examples include the Indus Waters Treaty between India and Pakistan and the Nile Basin Initiative, which includes countries in the Nile River Basin.

- **Regional Organizations:** Countries that share water resources can work together and coordinate more effectively. Organizations like the Mekong River Commission and the International Joint Commission for the Great Lakes facilitate discourse, collaborative management, and dispute resolution.

- **Global Water Initiatives:** International organizations and initiatives such as the United Nations Water, the World Water Council, and the Global Water Partnership all play important roles in improving global water security. These organizations promote knowledge sharing, capacity building, and policy advocacy to address global water issues.

Public Awareness and Education
Raising public awareness and promoting education about water issues is critical for creating a culture of conservation and sustainable use. Citizens who are informed and involved can raise awareness of the importance of appropriate water management and keep institutions accountable.

- **Educational Programs:** Including water education in school curricula helps teach a sense of responsibility and understanding about water conservation from a young age.

Programs should include information about the water cycle, water footprints, and sustainable activities.

- **Awareness Campaigns:** Public awareness campaigns can emphasize the need for water conservation and the consequences of individual actions. Campaigns can use media, social media, and community activities to reach a large audience and promote water-saving habits.

- **Community Engagement:** Involving communities in water management programs promotes ownership and accountability. Community-based projects, such as watershed management and local water monitoring, can encourage residents to have an active role in maintaining their water resources.

Technological Innovation and Research

Continued research and innovation are critical for developing new technologies and ways to combat the worldwide water challenge. Investing in research can result in discoveries that improve water efficiency, quality, and sustainability.

- **Research financing:** Governments, the corporate sector, and international organizations should all provide financing for water-related research and development. Prioritizing research fields like desalination, water recycling, and smart management systems can help promote technical progress.

- **Public-commercial Partnerships:** Collaborations between the public and commercial sectors, as well as academic institutions, can help to speed the development and deployment of new water technology. These collaborations can use resources, experience, and market access to scale up solutions.

- **Open Data and Knowledge Sharing:** Increasing open access to water data and research findings can foster collaboration and innovation.

Platforms that share data on water availability, quality, and usage can help guide decisions and inspire new solutions.

Creating a Sustainable Water Future.

The global water problem is a complex issue that necessitates immediate and concerted action. We can secure sustainable water use for future generations by comprehending the scope of the situation and putting innovative technology, laws, and practices in place.

Innovative water-saving technologies, such as enhanced irrigation systems, water recycling and reuse, desalination, and smart management, present viable alternatives for increasing water efficiency and reducing waste. To have a long-term impact, technology improvements must be supported by strong policies and governance frameworks.

Policies promoting sustainable water usage, such as integrated water resource management, water pricing, and effective regulation, lay the groundwork for equitable and efficient water management. International cooperation and public awareness are also crucial for tackling transboundary water challenges and instilling conservation values.

Individuals, communities, and nations have a shared duty to safeguard and manage our water resources properly. By embracing innovation, establishing sound regulations, and encouraging collaboration, we can overcome the global water issue and create a sustainable water future for all.

Chapter 7
Green Jobs and the New Economy: Creating a Sustainable Future.

The shift to a sustainable economy is not only an environmental imperative but also a chance to rethink our economic models. As the globe deals with climate change, resource depletion, and socioeconomic inequality, the concept of "green jobs" and sustainable corporate practices is gaining support. This chapter digs into the rapidly growing green economy, examining the rise of green jobs, the relevance of investing in sustainable firms, and the critical role of entrepreneurship in creating transformative change.

Green Jobs and The New Economy

Green employment, broadly defined, helps to preserve or restore environmental quality. They are fundamental to the new economy, which values sustainability, resilience, and social equality over traditional growth metrics.

Define Green Jobs.

Green occupations cover a wide range of industries and activities, including renewable energy, energy efficiency, sustainable agriculture, and waste management. These jobs are distinguished by their environmental benefits and contributions to long-term development.

- **Renewable Energy:** Jobs in renewable energy include solar, wind, biofuel, and hydropower. These positions include research and development, production, installation, and maintenance of renewable energy systems.

- **Energy Efficiency:** Energy efficiency jobs involve reducing energy usage through improved technologies and practices. This includes responsibilities for building retrofits, energy auditing, and the development of energy-efficient equipment and industrial processes.

- **Sustainable Agriculture:** Green jobs in agriculture require measures that improve soil health, limit chemical inputs, and increase biodiversity. This includes roles in organic farming, permaculture, agroforestry, and sustainable fishing.

- **Trash Management and Recycling:** This industry provides employment opportunities in trash reduction, recycling, composting, and the development of circular economy solutions. Workers in this sector create solutions that reduce trash, recover valuable materials, and convert waste into resources.

Green building and urban planning jobs are concerned with the design and construction of sustainable infrastructure. This includes architects, engineers, and planners who promote energy efficiency, sustainable materials, and green spaces in their designs.

The Economic Impact of Green Jobs.

Green employment stimulates economic growth and job creation while providing multiple advantages to society and the environment.

- **Job Creation:** The shift to a green economy is predicted to create millions of new jobs globally. According to the International Labour Organization, the transition to sustainable practices might provide

24 million jobs worldwide by 2030, compensating for job losses in traditional sectors.

- **Economic Resilience:** Green jobs help to build economic resilience by promoting industries that are less vulnerable to resource restrictions and environmental laws. Renewable energy, for example, reduces dependency on fossil fuels while mitigating the economic risks associated with fluctuating energy prices.

- **Social Equity:** Green jobs can help to improve social equity by creating opportunities for marginalized groups. Initiatives that prioritize skill development and fair employment practices ensure that the advantages of the green economy are widely distributed.

- **Environmental Benefits:** The fundamental purpose of green jobs is to minimize environmental damage. These vocations promote renewable energy, energy efficiency, and sustainable practices, which assist in minimizing climate change, reducing pollution, and protecting natural resources.

Investing in sustainable businesses. Investing in sustainable enterprises is an important part of developing a green economy. Sustainable businesses prioritize environmental, social, and governance (ESG) principles in their operations and decision-making processes, ensuring that revenue is balanced with beneficial societal effects.

The Rise of Sustainable Investment
Sustainable investing is gaining traction as investors appreciate the financial and ethical advantages of backing businesses that value sustainability.

- Environmental, social, and governance standards are used to assess an investment's sustainability and ethical impact. Environmental criteria measure a company's influence on the environment; social criteria look at its interactions with employees, customers, and communities; and governance criteria evaluate leadership, executive pay, audits, and shareholder rights.

- **Impact investment:** Unlike Environmental, social, and governance criteria, impact investment focuses on delivering measurable social and environmental benefits as well as financial gains. Investors actively look for firms and projects that address challenges like climate change, poverty, and inequality.

- **Green Bonds:** Green bonds are financial products designed expressly to fund initiatives that have a good environmental impact. These bonds are a popular way for governments, municipalities, and enterprises to raise funds for renewable energy, energy efficiency, and sustainable infrastructure projects.

- **Sustainable Funds:** Sustainable mutual funds and exchange-traded funds allow investors to invest in diverse portfolios of companies dedicated to sustainability. These funds provide an opportunity for private investors to help the green economy.

Benefits of Sustainable Investment

Sustainable investing not only promotes the transition to a green economy but also provides various benefits to investors.

- **Financial Performance:** Numerous studies have found that organizations with high ESG performance frequently beat their peers financially. Sustainable firms are better positioned to handle risks, recruit talent, and create long-term value, which leads to more stable and successful investments.

- **Risk Mitigation:** Investing in companies with strong sustainability practices can help reduce the risks associated with environmental legislation, resource scarcity, and societal unrest. Companies that proactively handle ESG issues are less likely to face fines, lawsuits, or reputational harm.

- **Positive Impact:** Sustainable investment allows investors to connect their financial objectives with their principles. Investors can help drive positive change by backing companies that care about the environment, social well-being, and ethical governance.

Challenges and Opportunities
While sustainable investing offers numerous benefits, it also faces

problems that must be overcome to reach its full potential.

- **Standardization:** Investors find it difficult to compare and evaluate sustainable investments because there are no established measures or reporting frameworks for ESG performance. Universal standard-setting efforts, such as the Global Reporting Initiative (GRI) and the Task Force on Climate-related Financial Disclosures, are crucial for increasing transparency and accountability.

- **Access to finance:** Small and medium-sized firms and startups sometimes struggle to obtain finance for long-term plans. Increasing access to financing through impact investing, and government assistance can help these enterprises grow.

Support for Green Entrepreneurship

Fostering a vibrant environment for green entrepreneurship necessitates supportive legislation, financing availability, and resources for innovation and expansion.

- Green business incubators and accelerators offer entrepreneurs guidance, resources, and financial options. These programs assist entrepreneurs in developing their business concepts, scaling their operations, and connecting with investors and industry partners.

- **Access to Financing:** Green entrepreneurs require funding to develop and commercialize their innovations. Venture capital, impact investing, and government subsidies can help fund early-stage and growth-stage firms.

- **Policy Support:** Governments can encourage green entrepreneurship by implementing policies that promote innovation and sustainability. This comprises tax breaks, subsidies, and regulatory frameworks to encourage renewable energy, energy efficiency, and sustainable activities.

- **Research and Development:** Investing in research and development is critical for advancing green technology. Research and development financing from both the public and private sectors can help to speed the development and commercialization of novel solutions.

Entrepreneurial Mindset and Social Impact.

Entrepreneurs with a sustainability focus frequently have a distinct perspective that stresses social and environmental impact alongside financial success.

- **Purpose-Driven Leadership:** Green entrepreneurs are driven by a sense of purpose and a desire to tackle global issues. This purpose-driven approach promotes innovation, resilience, and long-term thinking.

- **Collaboration and Partnerships:** Successful green entrepreneurs understand the value of collaboration and partnership. Working with other firms, governments, and non-profit groups allows them to increase their effect and achieve systemic change.

- **Adaptability and Resilience:** In today's fast-changing sustainability landscape, entrepreneurs must be both flexible and robust. Green entrepreneurs are often adept at negotiating uncertainty, changing their strategy, and devising innovative solutions to challenging issues.

Creating a Sustainable Economy
The green economy represents a fundamental change toward a more sustainable, equitable, and resilient future. Green jobs, sustainable enterprises, and entrepreneurial innovation are at the heart of this new economy, which provides prospects for economic growth, environmental protection, and social well-being.

Investing in sustainable firms and encouraging green entrepreneurship is crucial to accelerating the transition to a green economy. By prioritizing Environmental, social, and governance criteria, impact investment, and innovative solutions, investors and entrepreneurs may lead the way in tackling global concerns and producing long-term beneficial change.

Conclusion
Maintaining a Green Legacy

As we look to the future, the concept of a "green legacy" serves as both a source of hope and a call to action. This vision is more than just reducing the negative effects of human activity on the environment; it is an exciting plan for building a society in which sustainability, resilience, and equity are the foundations of our communities. In this final chapter, we will look at the vision for a green legacy, the critical relevance of hope and action, and the vital role that each of us plays in creating a sustainable future.

The Vision for A Green Legacy
A green legacy reflects a significant shift in our collective attitude, emphasizing environmental health and future generations' well-being

over short-term profits. This vision has numerous components, including environmental restoration, technical innovation, social equality, and economic reform.

Environmental Restoration

The rehabilitation and protection of nature are at the heart of a green legacy. This entails taking a comprehensive approach to environmental stewardship that addresses the interconnected issues of climate change, biodiversity loss, and resource depletion.

- **Climate Stability:** A green legacy envisions a world in which global temperature rise remains well below

2°C above pre-industrial levels, by the Paris Agreement. This necessitates a rapid shift to renewable energy sources, widespread adoption of energy-efficient technologies, and the deployment of natural remedies such as reforestation and wetlands restoration.

- **Biodiversity Protection:** Preserving and restoring biodiversity is critical to ecosystem resilience. A green legacy entails the creation and expansion of protected areas, the protection of vital habitats, and the promotion of sustainable land and water use practices.

- **Resource Regeneration:** Sustainable resource management is an essential component of a green legacy. Transitioning to a circular economy entails reducing waste, reusing and recycling commodities, and conserving natural resources. This change relies heavily on innovative approaches such as regenerative agriculture and sustainable forestry.

Technological Innovation

Innovation is the motor that propels the shift to a sustainable future. Technological developments in a variety of industries, including energy, transportation, agriculture, and manufacturing, are critical for decreasing environmental impact and increasing efficiency.

- **Renewable Energy:** The continuing development and deployment of renewable energy technologies like solar, wind, and bioenergy are vital to decarbonizing the global energy system. Innovations in energy storage and smart grid technology improve the reliability and integration of renewable energy sources.

- **Sustainable Transportation:** A green legacy envisions a transportation system that is efficient, accessible, and environmentally responsible. This includes the widespread use of electric vehicles, the development of high-speed rail and hyperloop networks, and the promotion of public transit and active transportation modes like as cycling and walking.

- Advanced Materials and Manufacturing: The adoption of advanced materials and environmentally friendly manufacturing methods can greatly reduce the environmental impact of industrial operations. Biodegradable polymers, sustainable textiles, and 3D printing are examples of innovations that show promise for a more sustainable industrial sector.

Social Equity and Inclusion

A truly sustainable future is egalitarian and inclusive, ensuring that all members of society benefit from sustainable practices.

- Just Transition develops legislation to solve environmental issues and promote sustainable development.

- **Grassroots Movements:** Joining or supporting grassroots movements and environmental organizations strengthens collective advocacy for change. These movements frequently address vital issues such as climate justice, conservation, and sustainable agriculture, and they rely on human support and engagement to make progress.

- **Public Awareness Campaigns:** Participating in or supporting public awareness campaigns educates others about the value of sustainability and inspires action. Sharing information on social media, attending community events, and engaging in creative activities such as painting and storytelling can all help to promote awareness and inspire change.

Participating in collective efforts
Collective action is critical for addressing the complex and linked challenges of sustainability. Working together in our communities, companies, and networks can have a significant and long-lasting impact.

- **Community efforts:** Participating in local sustainability efforts, such as community gardens, clean-up days, and energy-saving programs, creates a sense of belonging and responsibility. These initiatives not only address local environmental concerns but also promote social cohesion and resilience.

- **Workplace Sustainability:** Promoting and implementing sustainable practices in the workplace can have a big influence. Encouraging green policies, reducing waste, and boosting energy efficiency at work all help to improve the overall sustainability of businesses and organizations.

- Collaborative Projects: Working on collaborative projects with other stakeholders, such as businesses, non-profits, and government agencies, can help to increase efforts to address sustainability issues. Partnerships that combine various abilities, resources, and viewpoints are more likely to produce innovative and effective results.

Education and Skill Development
Lifelong learning and skill development are critical for meeting

the changing demands of the green economy. Individuals can make a bigger difference in sustainability initiatives by staying informed and learning new skills.

- **Environmental Education:** Studying environmental science, sustainability studies, and other related subjects provides individuals with the information and skills required to handle environmental issues. Formal education, online courses, and workshops give excellent learning possibilities.

- **Skill Development:** Learning practical skills in renewable energy, sustainable agriculture, and waste management improves our ability to adopt sustainable practices. Vocational training programs and certification courses provide opportunities for learning these abilities.

- **Leadership and Advocacy:** Developing leadership and advocacy abilities allows people to take an active role in driving sustainable efforts. Training programs in leadership, communication, and community organizing can help us lead and inspire others.

Personal Wellbeing and Sustainability

Personal well-being and sustainability are intricately linked. Taking care of our health enhances our ability to contribute to the health of the earth and society.

- Mindfulness and Mental Health: Practicing mindfulness and putting mental health first can help us stay focused, resilient, and motivated in our sustainability activities. Meditation, nature walks, and self-care routines improve general well-being.

- Healthy Lifestyles: Living a healthy and sustainable lifestyle, such as eating a plant-based diet, exercising regularly, and decreasing stress, is beneficial to both personal and environmental health. These decisions frequently accord with sustainable living concepts and might motivate others to do the same.

- Connecting with Nature: Spending time in nature promotes greater respect for the natural world and strengthens our resolve to safeguard it. Hiking, gardening, and outdoor recreation allow us to connect with and enjoy the environment we work so hard to protect.

The Path Forward

Building a sustainable future necessitates a shared vision, constant optimism, and consistent action. The green legacy we hope to establish is within reach, but it will require dedication and collaboration from all segments of society.

Our vision for a green legacy is a society in which environmental health, economic development, and social fairness coexist together. This vision is guided by the values of sustainability, innovation, and inclusivity.

- **Sustainability Principles:** Following sustainability principles ensures that our current actions do not jeopardize future generations' ability to meet their needs. This includes sustainable resource management, pollution reduction, and biodiversity conservation.

- Innovative Solutions: Embracing innovation in technology, business strategies, and governance is critical to meeting environmental concerns. By cultivating a culture of invention and continual development, we can create solutions that are both successful and scalable.

- Inclusivity and Equity: Ensuring that the benefits of sustainability are distributed equally is critical to a green legacy. This entails tackling socioeconomic disparities, creating opportunities for all, and empowering marginalized populations to engage in and benefit from sustainability efforts.

Unwavering Hope.
Hope is a powerful motivator that inspires us to envisage a better future and make the necessary efforts to make it happen. By having a positive attitude, we may motivate ourselves and others to stay committed to the path to sustainability.

- **applauding victories:** Recognizing and applauding victories, no matter how minor, keeps us motivated and strengthens our conviction in the possibility of positive change. Success tales remind us of our collective potential and the progress we've accomplished.

- Fostering Optimism: Maintaining an optimistic perspective allows us to be resilient in the face of adversity. Optimism inspires us to see opportunities where others perceive difficulties and to persevere in our attempts to create a sustainable future.

- Creating Community: Surrounding ourselves with like-minded people and supportive communities fortifies our determination. Building networks of people who share our vision and beliefs fosters a sense of unity and shared purpose.

Continuous Action

The sustainability journey is continual and requires consistent work. Continuous action, directed by our vision and motivated by optimism, is required to make genuine progress.

- **Setting Goals:** Setting specific and attainable goals allows us to stay focused and track our progress. Setting sustainability goals provides guidance and motivation at the individual, community, and organizational levels.

- **Adapting and Learning:** The path to sustainability is dynamic, and we may need to adjust and learn as we go. Being open to new information, willing to change direction, and devoted to lifelong learning are all essential for being productive.

- **Taking Responsibility:** We all have a responsibility to contribute to sustainability in whatever way we can. By accepting responsibility for our actions and their consequences, we can make informed decisions that support a green legacy.

As we end our investigation of creating a sustainable planet, it is evident that the path to a green legacy is both difficult and rewarding. The shift to a more sustainable future necessitates collaborative effort, innovative solutions, and a strong commitment to environmental stewardship and social equality.

The aim of a green legacy is more than an aspirational dream; it is a practical and feasible goal. By embracing hope and taking consistent action, we can overcome the problems we confront and build a society where sustainability is the norm rather than the exception.

Your involvement in creating a sustainable future is vital. Every activity, no matter how tiny, adds to the bigger drive toward sustainability. Making mindful decisions, advocating for change, and joining in communal activities may all contribute to a greener, healthier, and more equitable world.

Together, we can create a green legacy that honors the past, fulfills the demands of the present, and secures a prosperous future for generations to come.

www.ingramcontent.com/pod-product-compliance
Lightning Source LLC
Chambersburg PA
CBHW071206240526
45470CB00018B/1518